EINSTE

For st1

Einstein - Rosen Bridge Theory Now Proven: Wormhole Allows Entire Person to Travel

*Applied Physicist
Johnny Vincento*

JOHNNY VINCENTO

Einstein - Rosen Bridge Theory Now Proven: Wormhole Allows Entire Person to Travel

By Johnny Vincento

UNITED STATES COPYRIGHT OFFICE

Library of Congress

ISBN: 9798643293590

FIRST EDITION 2005

SECOND EDITION 2020

Johnny.vincento@yahoo.com

All rights reserved, Email me to request permission to use any material within this work. Have exactly what subject material you are requesting in your Email.

COVER IS NIELS BOHR (Left) AND ALBERT EINSTEIN (Right)

ABSTRACT:

What has been done?

For the last twenty years, Teleportation has been successfully accomplished. It is not one's body that Teleports, but the life force within one's body. Harnessing one's own life force opens a mini individual Wormhole. Thus, correcting the one way Einstein - Rosen bridge theory into a two way working Wormhole. This idea was brought about by Nathan Rosen himself.

What are the main findings?

By using a Formula, one's body is weakened to the point of opening a Silver Cord.

What is a Silver Cord?

The Silver Cord is a natural law of physics that creates a Wormhole that one's Life Force opens. The Silver Cord is a tube that bends Space and Time. Yes, even Time itself. All healthy people can harness their Silver Cord. One's life force is the only thing that can cross dimensions and rematerialize a second physical body. Furthurmore, Space is not travelled, but bent, so there is practically no distance to travel. For Point A is folded on top of Point B.

What Physics are used in this find?

One's physical body is here on Earth, while through Quantum Entanglement and Spooky action at a distance, produces another duplicate of that entire physical person as one's life force enters. The Silver Cord is invisible for the most part and connects to Multiverses and the Mutiworlds in those Universes.

INTRODUCTION:

Twenty years of actual Teleportation shows this work is also consistent with The General theory of Relativity, Standard Model, (As far as the particles), Multiverse theory (Now proven) and String theory (In reference to creating something from nothing).

However, Quantum Entanglement easily accounts for that since we are made of particles that can be in more than two places at once.

The Einstein - Rosen bridge, Quantum Entanglement, General theory of Relativity and Spooky action at a Distance, all goes in a beautiful harmonic balance with this Working Teleportation and Wormhole discovery.

There may still be some disagreement with how Quantum Entanglement can influence at a distance. This being the particles of (Person A) connected by Spooky action at a

Distance to (Person B). This disagreement is: Does (Person B) become an exact physical copy or an opposite mirror copy? Both are satisfied, for the copy is exact and at the same time opposite: (Person A) on Earth is mortal and (Person B) across dimensions is immortal. Both the same person, however, one's life force can only be in one body at a time.

How do Quantum Particles then communicate?

They could be explained if they were sending signals. One particle telling the other particle what properties to have. And this is where Albert Einstein and Niels Bohr separated basicly the whole science community. In order to have a particle in another galaxy and a particle here on Earth for example: The signals would have to travel faster than light. Einstein's General theory of Relativity won't allow that. Bohr's Quantum Entanglement theory said it could be done, but couldn't prove it and Einstein said that it couldn't be done. So there you have it for about one hundred years: Quantum Mechanics on one side and General theory of Relativity on the other: Separtated.

Now comes this research that can be proved. This can be done easily with a study group. Proving without a doubt that the time has come for anyone to test and see with their own eyes. Even filming a Teleportation may capture something as well. No one can deny seeing with their own eyes and experiencing Teleportation. Many older people will have a hard time doing the Formula. The test group should be for healthy people 28 years old to mid thirties. The older you get the harder it is to Teleport. That age range is when I had the best results.

What are the limitations of opening a Silver Cord Wormhole?

It is the time limit. It will stay open for no more than five minutes. Then a person will come back whether they want to stay there or not. I wanted to stay once and the Wormhole vacuumed me back.

The other limitation that is now solved was Steering the Wormhole to a specific location. For almost the entire twenty years of Teleporting there wasn't any Steering. I would just end up where it took me. That Steering issue is now solved and the potential is endless.

What are the future goals and what is hoped to be achieved?

Now that the Steering is very exact all things can be tried and tested in a scientific manner. One can even Teleport to visit a specific person, place or other world across dimensions. Even time travel can be tested. **One can go back in time** as you will see. However limited success, it shows that it can be done. There is a lot of Teleportations that will be exciting for science to attempt: Going to Mars and activating the Formula. What will one find when they walk on an immortal Mars the way it was: If there is a duplicate overlapping Mars? Teleporting back in time to see an event happen. Steering the Wormhole to visit a planet with simular intelligent life besides humans.

Teleporting while in Pompeii. That whole city was buried in about a day. Is the entire city still booming across dimensions?

The distance from the overlapping world closest to us, occupying the same space, is about the thickness of a hair or at most the side of a coin. This is no guess either. That is the veil thickness, no more than that. I was there many times and was in the body that overlapped my body. One body was in my room, the other body was in the room that overlapped my room. My bodies didn't move. Same body locations, same rooms, different furniture for the most part, different drapes, different paint. Yet the same location in Space and Time.

The greatest goal would be to benefit humanity with obtaining Anti Gravity technology. Not to mention, the knowledge to build a more stable communication between at least the closest overlapping world. This would be the technical support to build a **"Dimension Separation Viewer"** to have a window between worlds. It is possible to create, because the atmosphere lifts the veil between worlds on accident at times. This random event can be duplicated in a lab environment. The main application to make such a machine would be "Vibrational Waves." Furthermore, from my experience, building the "DSV" would also require wearing torch glasses or very heavy sunglasses. Light and matter are different across dimensions. When one Teleports, the physical body that is materialized and duplicated is immortal. The natural laws of physics during the Wormhole and Quantum Entanglement materialization to the second body: Aligns within that worlds "Vibrational Matter" and "Light." That is the reason that "Vibrational waves" are needed to see the different "Vibrational Matter." which occupies the same space.

Light is not the same and matter is not the same as here. The Teleportation is safe everytime, but if a "DSV" machine is built our bodies won't be aligned with that worlds light of different photon wave particles and frequencies. The photon particles of an overlapping worlds light are in tune with that worlds matter. The light from that world won't affect the person Teleporting due to being in a rematerialized physical body of that matter in tune to its worlds light. However, if a Wormhole was to open and it did on me once: I guess someone wanted to contact me with their own DSV. The light was so bright that if I would have looked at it I may have been blinded.

One other note: If one is paralyzed and Teleports that person will be in their fully functional immortal copy. There are some ways of trying to bring that healing back, but it would take Teleportation of that person to explore any options. The person would walk in their new body for the approx. five minute Teleport. I witnessed this so that is why it is a fact.

METHODS:

Attain the "Teleportation" in the Summer on a Sunday

FORMULA FOR THE QUANTUM ENTANGEMENT TELEPORTATION
This is the THREE DAY PLAN
for those who are thin or petite:
9am Thursday – 9am Friday: Fast and sleep.
9am Friday – 9am Saturday: Fast and sleep.
9am Saturday – 9am Sunday: Fast and do not sleep.
At 8am on Sunday sip on 80 proof
liquor to obtain great
fatigue only.
You can just do extreme exercise if
you wish not to use 80 proof liquor for
fatigue. For example, jogging in place
until you are extremely exhausted.
At 9am close your eyes and sleep: immediately
you will be in your resurrected body of flesh
in one of the overlapping physical worlds.

FORMULA FOR THE QUANTUM ENTANGELMENT TELEPORTATION
This is the FOUR DAY PLAN
for those who are of average weight:
Same as above except add:
9am Wednesday – 9am Thursday: Fast and sleep.

The body must be made weak enough to visit the other worlds. You only have up to 5 minutes there, so it is a recommended idea to have a friend wake you after exactly 4 minutes and 45 seconds. This is so you do not feel as you will be stranded there. **Also, do not have your friend touch you around your mid section: There is an invisible "Silver Cord" which is what the "Wormhole" is. It is attached to both physical worlds and both physical bodies. It is attached to round about the center of your chest.**

This formula must be adapted to your individual body. It is very close, but not exact. If you activated the formula and did not receive the Journey, then the next day repeat the fasting and fatigue: That should accomplish it. Read "Journey to my Father," I lost my nerve and no longer wanted to take the Journey: But it was too late, the formula was already activated.

WHEN YOU COME BACK FROM YOUR JOURNEY START TO EAT SLOW. FOR

JOHNNY VINCENTO

EXAMPLE A LITTLE EVERY HOUR. BY THE END OF THE DAY YOUR SYSTEM WILL BE NORMAL.

RESULTS:

The following Journeys through the Silver Cord Wormhole are my Reports of Exploration. This is a **short compilation** of the many. They are to be analysed in a scientific investigation. As will all other Einstien - Rosen bridge Journeys, when other test groups are formed. These deciphering of the Reports will allow for great knowledge into the possibilites of crossing Time, Space, Dimensions and Multiverses. It was written long ago before human memory, going back so far that estimates can't even date or comprehend the tens of thousands of years the stone writings were made: One says: "nine are the worlds within worlds" I visited four.

JOURNEY
TO ANOTHER WORLD THROUGH
"THE SILVER CORD WORMHOLE">
JOURNEY TO MY FATHER

Fasting for three days, I sipped on 80 proof liquor for courage. For I know, the time has arrived for my trip to the unknown world. And courage is indeed needed, for the Journey is taken only by the boldest of individuals. With liquor providing courage, it also provides the needed fatigue that is combined with the fasting on the third day. This accesses the secret of the world: <u>For the world is worlds within worlds</u>.

I fell asleep at some time past noon and this was a Sunday.

Immediately I was standing in my physical Quantum Entangled duplicated body, in the home that occupied the same space as mine: In the other world. Before me was a sight not to be taken easily:

I entered the last room to my left and seen my own Father: Looking through chest high stacked cardboard boxes. It was good to see a friendly face. I went to the window of the empty room and looked out. The second story view was strange to me, since I knew this home occupied the same space. How come the street cul-de-sac was the same shape, but covered in crushed white rocks? In my world the street is blacktopped. How come there is a horse corral across the street? There is only fields without a home in sight? In my world all these fields are full of homes! <u>THEN I SPOKE TO THE ONE WHO WAS IN THE LIVING DUPLICATE IMAGE OF MY OWN FATHER ON MY EARTH.</u>

This is what was said:

"Father, where am I?"

"You are here."

"No I'm not, I'm at home sleeping."

"How could you be at home sleeping if you are here?"

"Listen, I'm telling you, I'm not here, I'm at home sleeping in bed."

"Cannot you understand the difference, between being at home sleeping in bed and standing here,

in front of me, in this room?"

"I don't care what you say, I need to wake up, I want you to slap me in the face!"

"Are you sure?"

"Yes."

(Then I was slapped front hand and back hand)

"Harder again!"

(Then I was slapped the same, with no pain, only impact pressure)

"I need to get out of this house!"

"You will never make it past the people in the hallway."

"Do you have any weapons?"

"No."

"Not even a kitchen knife to open your boxes?"

"No, I do not have any weapons."

"Well I'll take my chances!"

Once my Father said I would never make it past the people, it was clear to me that this was not my Father and a trick of that world. For he did not sound the same or use the same vocabulary. For I thought at that time my Father is alive in my world: How can he be here too? These are the things I thought at that time, for I did not understand yet, that I have two Fathers: One on Earth and one in that world.

I then thought to myself I'll rush the hallway while their backs are turned, then turn around after passing them, so they can't jump on my back. I charged the hallway expecting a fight, I turned to face them and I was walking backwards. Their walking shuffle was getting faster. I visualized the kitchen knife I wished I had because they were closing in. Then the knife I thought of appeared out of thin air, on the hall room white rug. I picked it up and it was tangible and made of stainless steel. I could feel the steel. I had a hard time telling the difference between this world and that one. The worlds are exactly the same physical wise: **You can't even tell the difference!** I ran down the stairs. I could hear my feet banging down the stairs. The floor plan

JOHNNY VINCENTO

was the same, so I knew how to get out of the house. I grabbed the brass push button handle and rushed out of the house.

I then tripped on high ground during my mad rush out and fell in the grass. The grass was the greenest grass that I have ever seen: I did not see one dead blade, the grass was about five inches high. I could feel the grass and then I rolled over on my back. I stretched my arms out and became intoxicated with pleasure, with the sight of the sky and clouds on that beautiful day. It was as if my emotions were multiplied times ten. I thought to myself:

"This is the first time I have ever been outside in this world. I want to see how far this world goes beyond the house itself."

I thought if I turn around and see the horse corral across the street, that will prove that the sight from the upstairs window was not a false one. I turned around and seen a joyfull sight: The brown horse, stone road, wooden fence were all there, just like I seen from the window.

I said out loud:

"This isn't just a house, it's a whole other world!"

I got up and stood in the grass facing the house and it looked different from my world.

I yelled:

"Father get down here now, you got some explaining to do!"

He said:

"I will be down in five minutes."

I yelled out:

"Five minutes is too long, I want you down here in 30 seconds!"

He replied:

"Ok"

and I awoke in my bed, lying on my side. The weather too was beautiful in both worlds that day.

DISCUSSION:

Is there some way that Quantum Entanglement made a physical copy of my

JOHNNY VINCENTO

Dad on Earth? This is a copy of him in physical form, but not connected by Spooky action at a distance. **A copy made by Quantum Entanglement of the same age, yet acting independantly** from the first "Particle" so to say. **No communication yet duplicated at one time.** How could there be two of him? He had a different personality and Life Force. This is written in ancient times as well: "When you see the one not born of woman...that one is your father." The sky had white clouds, my voice had sound, (That implies that there is air) the home was the home over the home, occupying the same space. The room my Dad was in was full of auto parts in this world, over there just a wall of boxes single row. The room had blinds in this world. Over there it had drapes. The street was the same shape and location, but covered in rocks. No other homes were seen around. When Equations are made take into account how these things can be. The **String theory** and **Quantum Mechanics** take over with the thinking and visualizing of the knife. It was just as I had in my mind. A hollow stainless steel butter knife that was in our kitchen. It even had the molded handle and spiral decor. It was solid and it was real metal. I picked it up in amazement. The mind creates whatever one wishes for it seems, like a transmitter creator. I would even add my **Theory of Matter** to combine it with **the String theory. Which states: That in other overlapping worlds the entire atmospheres are full of invisible building blocks of creation. This theory also goes into the possibility that individual galaxies didn't form from a Big Bang theory, instead, through the new "Daisy Field theory" and those galaxies are formed from White Holes from an overlapping universe.**

The Deep Field Hubble telescope images show what looks like a daisy field and the Black Holes in the center of each galaxy used to be a possible White Hole. **In those photos it looks like fields of daisies, but each daisy is an entire galaxy.**

RESULTS:

Same house as previous. This world is the closest overlapping Universe or plane of existence.

JOURNEY
TO ANOTHER WORLD THROUGH
"THE SILVER CORD WORMHOLE">
THE BIRTH OF THE SPIRITNAUT

Alone with my spirit, equal was the formula as the last. Beautiful visions of which cannot be believed. It's time to see if the formula is perfected. Three days of fasting and my body is weak enough to send my Life Force. I sip slowly on 80 proof liquor, I pray to see all things unseen. The liquor does its purpose. Sleep takes hold heavy, my eyes heavy, I say last out loud, **"Spirit take me to the other world."**

Sitting half upright with my arms crossed over my chest, I close my eyes to find myself immediately in my Quantum Entangled duplicated body. I was standing in a room, with a window before me, and a desk to my left. I say to myself out loud:

"I made it, this is great, I am really here!"

My joy was times ten. I patted my sleeveless arms and could feel my flesh. I then patted the front of my legs. I was wearing the same checkered cloth pajamas. They were duplicated with my body. I said to myself

"I don't want any doubt what so ever,"

that is when I snapped the elastic on my pajamas to my skin. I then said out loud,

"Yeap, I'm here."

I felt the white see through drapes and it was made of cloth. I rubbed it in my fingers. I pulled back the drapes and I was looking down from the second story of a residence. The street was new black top and it was lined with new round top trees. There was a sidewalk along the street and a sidewalk to the home I was in. There were homes on the other side of the street. It was night out and I seen no street lights. The other or same overlapping moon must have given the dimness.

I then turned around to explore the room and I seen a kid with blonde hair of the age of about 17. It was his room because he was sitting to the left of the room. Before me straight ahead was a single door. He said to me

"What am I doing in your dream?"

I said,

"This is no dream, I have been practicing spirit surfing. I don't know how much time I have, so what's your name?"

I shook his hand and he was real, because the flesh was real. I will disclose his name, even though I will be bombarded with inquiries from the whole country. He replied,

"My name is Bates."

It could be spelled differently.

I was then vacuumed through a tube and I could hear the **"Whoosh."** Everything went black, however I was fully conscious. I started to panic, feeling that I would not make it back to my world. So for that three or four seconds I prayed:

"Please help me get back to my room, please help me get back to my bed."

I then arrived in my bed sitting half upright with my arms folded at my chest. I opened my eyes to a sight that was not to be taken easily: **It was not my room!** The walls were hunter green, mine are white, and the furniture was different. However, my king size bed was in the same position as was my body. I said with my voice shaking

"This isn't my room."

Then I was vacuumed through the tube again. I heard the same **"Whoosh"** and I then opened my eyes to the same bed same positioned body, but the correct room.

. . . .

DISCUSSION:

Later after other visits, I discovered the house I visited, was the home occupying the same space, but in the second world. The room with the kid was the one across the hall on the East side of the home. The floor plans are the same. **Then the Wormhole disintegrated my Quantum Entangled body and vacummed me into it and brought me across the hall to my bedroom, right through the walls.** I had my thoughts and mind. I was looking at the side of the tube which was black. I never had consciousness before in the Silver Cord Wormhole. I wonder if I turned somehow to the front, if I would have seen the opening. Then the Wormhole rematerialized my body again, but it was the bedroom that overlapped mine in the second world. Did my Life Force Silver Cord make a mistake? Then I spoke aloud and corrected the mistake. Both bodies were in the same location: One mortal the other immortal occupying the exact same space. So not a big mistake. Or was I brought to that situation so I could gain knowledge by showing me that?

NOTES AND EQUATIONS

JOURNEY
TO ANOTHER WORLD THROUGH
"THE SILVER CORD WORMHOLE">
A CREATURE EXTINCT IS ALIVE

After the third day of fasting, night began to fall. I prepared to engage the final step of my "Silver Cord." As I sipped on my liquor, I readied myself to harness the powers of the Universe; and to walk within the parallel dual Earth as was written. The affects of the liquor come, but only severe fatigue is what I seek. Sleep arrives moments after I set down my 80 proof glass.

I find myself standing in my Quantum Entangled duplicated body. In a home well decorated, about 20 miles distance. The name of the city in this world is River Forest. I was visiting the city that occupied the same space, so I don't know if they named it the same. The names of the streets were Franklin and Greenfield in this world, however it's the same space in that world.

It was two homes to make one. The corner North East combined with the home of

North. Walking through a room, there were a very many people around me, but not close to me. Before me was a fancy wooden coffee table, with a toy robot dog, made of plastic or aluminum. The color was gun blue and silver. It was about two feet long with four legs about a foot long each. It was about 8 inches wide. The only thing that was familiar with the toy, was its gun blue visor, covering the eyes. I seen those on toy robot dogs in this world, those toys are only about 6 inches high and 8 inches long. This toy was huge. **I purposely went on this Journey to find information.** I picked up this toy looking for a manufactures label. I turned it upside down and seen a blank square white label.

I walked on to cross the addition on the second floor that connects the two homes. The West side was all windows overlooking Franklin Avenue and the homes across the street. To my left were some dorm size rooms. One guy, about the age of 25, with black hair, was sitting in his room in the doorway on a black cloth chair. I asked,

"All of you couldn't possibly be related, there are

too many, do you all live in this home?"

He said,

"No one here is related, but we all live here and get along fine."

Stepping out of the overpass addition, I entered into the second home. Before me were two women with a creature that I have never seen: And the creature was alive! It was the color of dark olive green. I petted its head and its flesh felt like puddy. The living thing appeared to be a dinosaur. Sitting on rear two and standing on front two, its body measured at least two and a half feet high while upright – about two and a half feet wide – three feet long (without neck or tail) – the tail, about three feet long and pointed at end. While upright, its curved neck rose to my chest with its head easy to pet. The curved neck if straightened would measure about three feet long. Its head was about a foot or 11 inches wide, with some type of reptile lips. As for the description of the skin, it was one solid color and its texture was smooth like clay. The spine could be seen on its oval back. The snout of the animal was almost flat with the slightest of an arch, as was the top of its head, but rounding at the sides. The depth of the snout is unknown, for I was looking at it straight on. However, a guess of 4 to 6 inches would be round about. Its eyes were centered for front viewing, not like illustrations where the eyes are on the side of the head. As for the appearance of the eyes, they were not like a reptiles, nor as a birds, or a cats, instead they were like a puppies and pure goodness emitted from its eyes and could be felt. For this reason, it had a spirit similar to a dogs and was above a reptile. <u>It resembled a Brontosaurus and the people had it as a pet</u>.

With the quick sound of suction, the creature swallowed half of my right arm. I felt only pressure, no pain and I had no fear because pure goodness was within the living things eyes. The women started laughing and the one in the chair said

"Don't worry, he always does that."

The creature looked at me with its head tilted to my right and gave me a look as if saying

"Are you going to stay?"

Then the immortal creature let me pull my arm out and I returned to this world.

DISCUSSION:

The homes that were connected are not connected here. Seems that this was the closest overlapping world.

RESULTS:

JOURNEY
TO THE SECOND WORLD THROUGH
"THE SILVER CORD WORMHOLE">
CHILDREN OF HEAVEN

In the year 2005 on the 15th day of May, I closed my eyes for the time was here. This was a Sunday at exactly 8:30 am.

I found myself standing in my Quantum Entangled duplicated body, in a field with a large tree to my left. I felt my legs, arms and put both hands on my face. I thought to myself,

"I'm here again."

The worlds are exactly equal physical wise and conscious wise.

About 35ft ahead of me was a single story home with a wooden outdoor deck. The deck had a balcony around it. There were a very many people as if at a get together. I walked into the gathering and seen all the kids were on the deck, which was about two steps off the ground. The color of the deck was unfinished pine looking wood. The deck was attached to the left side of the home.

As I turned left around the deck to the front of the house, two kids approached me. One was black haired about the age of ten. The other was about 13yrs old with blonde hair and taller. This is what happened:

"Do you want to see something?"
"Sure what?"

They pointed down between them and there was a round ball, the size of a volleyball. It was the color of liquid water. I could see the white pavement through the water. The first 10% was like the color of pale flesh, however, the correct color is not a color at all. I can only describe it as looking through hot light soaped water, into a white porcelain bath tub. Now this ball was a spirit and moved as the kids moved. I said:

"Yeah he's really cute."

I could not see the face of the Life Force. I was looking down on it as it hovered a bit off the pavement. And even when I did see the part where the face should be, this being later before I left, it did not have a face. It was a round spot which was the color that I am trying to describe. It resembled a circle within the sphere of liquid water. I got the feeling that this ball was the spirit of a child, possibly one not even born yet.

I then seen my Father walking onto the front of the driveway. My spirit rejoiced while I walked up to him. He took my left hand with his right and we walked to the home. The rejoice was short lived. At the front of the home I stopped and pulled his hand back. I said:

"Are you my Father?"

Telepathically, within my head he said:

"What, you don't know who I am?"

I panicked again and started yelling to the people around me:

"Where am I?"

No one would talk to me as if they were not suppose to give me any information. However, the blonde 13 year old kid leaned down toward my ear, from standing on the first stair of the balcony and said very quietly:

"Everything is true."

Then the two started talking among themselves about if I was one of them. While they talked I looked over the many people which was over fourteen. I recall the 10 year old very specifically within their conversation saying:

"Maybe he's not one of us."

After their debate, the 10 year old said to me very straight out:

"Are you a nighty?"
"I don't know, what's a nighty?"
"They usually come to visit at night time and then they are picked up. Are you going to get picked up?"
"I hope so, how about you, are you ever going to get picked up?"
"No ------- I'm never going to get picked up." (And he said that with disappointment)

I then returned and opened my eyes to my world and the weather was the same.

DISCUSSION:

The kids knew I was visiting through Teleportation. did you read what they said? **"They usually visit at night time and then they are picked up."**

RESULTS:

JOURNEY
TO THE SECOND WORLD THROUGH
"THE SILVER CORD WORMHOLE">
AERO DYNAMICS

Those of you who are students of **New Laws of Physics:**

I will briefly explain, I Teleported and found myself in my Quantum Entangled duplicated body in front of the warehouse that overlapped my families business in the other world. I patted myself and felt my thighs and arms. Both worlds are equal, one cannot even tell the difference physical wise or conscious wise. I went into the front office area and then into the kitchen dining area. There were no lights, only numerous candles. I met a guy who was solid. We were sitting at a rectangular wooden table. He was facing North, I to the South. To my right, touching the wall, was a stack of white standard looking paper. The size was that of a stack of about 250 sheet block, like you see at the store. He said to me

"Watch this"

and he pointed his right index finger at the stack and made a single sheet of paper hover about a foot off the table. Then he quickly moved his finger to his right and the single sheet of paper shot across the room straight like a dart. I then heard it hit the wall with force. I said

"That's neat do it again."

He then did the exact same thing.

DISCUSSION:

Maybe your equations can explain why the pieces of paper acted that way. I went there on another Teleportation journey as well. Same place and the family business had new looking sheet metal sides of the building: That corrugated type sheet metal. Over here the sides of the building are old and rusty.

RESULTS:

JOURNEY
TO ANOTHER WORLD THROUGH
"THE SILVER CORD WORMHOLE">
FATHER HOW LONG WILL I LIVE?

In the year 2006, on the 26th day, of the third month, I was two full days fasting past. This was a Sunday, in the early dim hours. I did not expect the Journey so early, but it goes to say I was very weak.

I found myself in a courtyard in my Quantum Entangled duplicated body. The area was surrounded by a wooden fence. The fence was four inch boards standing side by side and each was rounded at the top. The sky was just starting to become a dim white. I yelled out:

"Father where are you?"

I seen him very quickly. He appeared in the image of my own father on Earth and was in the flesh for I shook his hand. This is what was said:

"Father it is good to see you again. I took great effort to see you, but our time is short together. I came to ask you some things."
"What do you want to ask me?"

"I want to see the future."

"Not everyone has the ability to see the future."

"I want you to show me the future."

"Well let's see if you possess the gift."

"I want you to concentrate on where I am pointing and think of seeing the future."

"You mean between those two trees over there?"

"Yes, do you see any visions of the future?"

"No."

"Let's see if you can see the past. Concentrate on

where I am pointing and think of Indians, do you see any visions?"

"No."

"Then you do not possess the ability."

"Father how long will I live?"

"To seventeen."

"I am much older than seventeen."

"Everyone has an inner age."

"Like an emotional age?"

"Yes."

"But I want to know how much longer I will live?"

"Don't worry, you have many numbers."

Then we walked into a department store type setting. The courtyard was connected to the building. There was a great amount of clothes on round chrome racks. I walked to a room that had no door, and seen a woman washing her hands in a sink of about 60's in age. I said:

"Excuse me, how old are you?" and she replied slowly **"Fifty thousand years."**

I then wanted to leave and yelled out:

"Father where are you?"

I turned around and I seen him about ten feet away, went up to him and put my left hand on his right back shoulder and said:

"Get me back, I want to go back right now!"

Then I returned to this world.

DISCUSSION:

As you read the conversation with my Father in that world that is how fast it was, so to say, in real time. I did not have time to concentrate on seeing the future. The only thought I had when he was pointing, was that I want to see the future. Concerning the attempt to see the past, I did see the past. It was however, only for a quarter of a second. I seen my body, lying on my right side on a floor, I could see my front thighs and I was wearing navy blue sweat pant type material.

I was looking at an Indian toy that I received from an island in Lake Michigan when I was about 5 or six years old. The toy was a painted wood carved totem pole. This toy is what I was thinking about while he was pointing. And after some years I understand why he said **"and think of Indians."**

HOW TO STEER:

One must think of something very specific to see the Past or the Future. And this is how you steer, not just in Time, but across dimensions as well.

Two weeks before the Teleportation, one must tell their Life Force where one wants to go in exact detail. This will be done everyday at least twice a day. Then when the formula is complete and one closes their eyes, their life force will take that one there. This is a fact for crossing dimensions and crossing time. However, when I travelled time I did it from the other side of Dimensions. This must be tested to see if time travel is possible from our mortal plane.

That is if one has the ability to see the future. As far as my vision, it was complete, as if I was there. Everything with my Father happened so fast and the vision was not even a second long. If I could have concentrated, I would have been able to see more: At least for the past. As far as the future goes, I did not concentrate.

Now that the knowledge of how to Steer the Wormhole is understood, Not asking for help and directly taking the trip: **Steering to the Past or Future can be tested. An exact day, time, place or even person must be steered to.**

This would not be across dimensions as compared to the other Teleportations to planets and Multiverses across dimensions, but would be **an attempt to travel through time on our Time plane of existence**: Past or Future can be tested.

The Quantum Entanglement was pushed to the limit in this Journey. Being in three places at once.

First, I was on Earth in my mortal body.

Second, I was duplicated across dimensions in my immortal body.

Third, From my immortal body I was taken back in time to when I was a child looking out of my own eyes.

See three places at once.

Niels Bohr said:

"Everything we call real is made of things that cannot be regarded as real.

If Quantum Mechanics hasn't profoundly shocked you, you haven't understood it yet."

"

CONCLUSION: This is a great amount of knowledge obtained and I am only one person. The time travel possibility is very exciting for the future of Humanity and Science.

It has been around one hundred years now. Finally, The General Theory of Relativity and Quantum Mechanics are unified. I think Bohr, Einstein and Rosen would be happy.

REFERENCES:

"The Theory of Special Relativity"
Albert Einstein published 1905

"The General theory of Relativity"
Albert Einstein published 1915

"Can Quantum Reality be considered complete?"
(The Einstein - Podolsky - Rosen Paradox)
Albert Einstein, Boris Podolsky, Nathan Rosen
May 04, 1935

JOHNNY VINCENTO

"The Particle problem in the General theory of Relativity." A. Einstein and N. Rosen Institute for Advanced Study, Princeton (received May 08, 1935

"How to actually Teleport through dimensions and visit other worlds" J. Vincento 2005 2019

NOTES AND EQUATIONS

JOHNNY VINCENTO

NOTES AND EQUATIONS

NEED TO KNOW THE TRUTH ABOUT EXISTENCE?

Go to AMAZON
type
JOHNNY VINCENTO

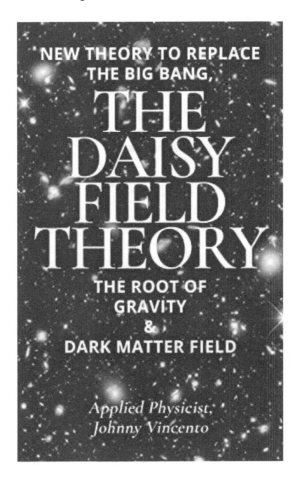

JOHNNY VINCENTO

GET THIS MIND BLOWING SERIES:

GO TO AMAZON

TYPE

TELEPORT NEWS

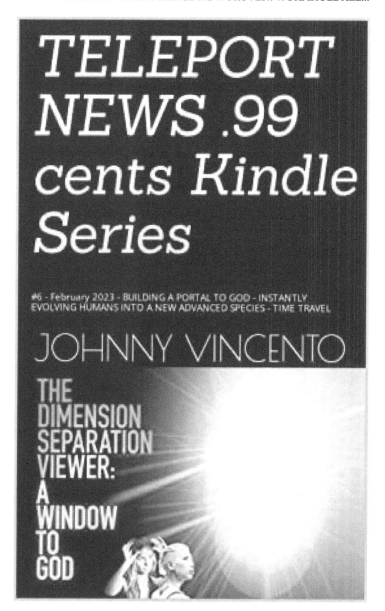

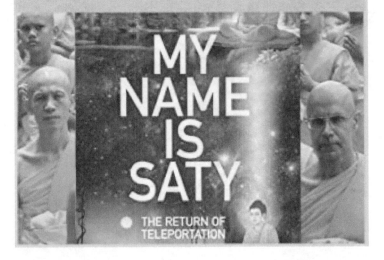

EINSTEIN - ROSEN BRIDGE NOW PROVEN: WORMHOLE ALL...

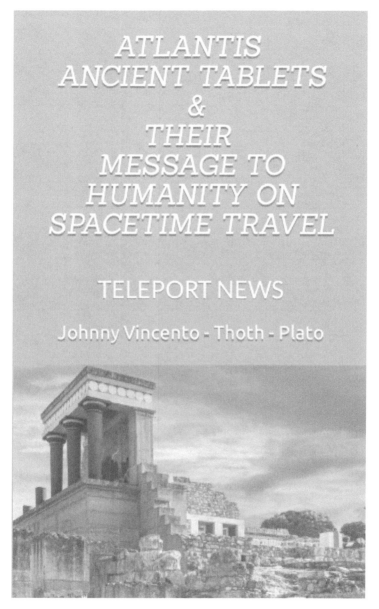

AVAILABLE IN PAPERBACK HARDCOVER KINDLE

JOHNNY VINCENTO

TELEPORT NEWS, brings you a rare paperback edition. Taking you right into lost eyewitness accounts of Atlantis. It describes the secrets of who built the Sphinx, Great pyramid and what they were built for. The science formula deciphered from these tablets has been experimented on for 20 years with great success. They were written ages ago in stone, estimated to be around 12,500 years old.

The writings were discovered and never understood since around 750 AD. They are written from an extremely advanced ancient scientist. As the world leading physicist in full person teleportation, I was able to decipher them. Only someone familar with these applied forces of quantum entanglement and wormhole travel to interdimensional parallel worlds, would understand what he is saying. I can see why no one understood them. The Project actually did exactly what was written and the findings are mind blowing. The ancient scientist said that he purposely hid the tablets for the future of mankind and what a story he has to tell us!!!

Johnny Vincento is a physicist in the Applied field of Quantum Entanglement Wormhole travel. He says "I'm not a pastor or a preacher, I am a scientist & messenger to the entire world on what the TELEPORTATION PROJECT discovered. I'm one of the few ever born to see Earth's exisitence from outside it's existence."

ISBN 9798355503369

AVAILABLE IN COLOR PAPERBACK KINDLE

JOHNNY VINCENTO

GRAB YOUR COMPASS AND SHOVELS AND GET READY FOR THE TREASURE HUNT OF ALL TIME!!! ONE THAT WILL PROPEL HUMANITY THOUSANDS OF YEARS INTO THE FUTURE · THE LOST LIBRARY OF THOTH - CONTAINING OVER 36,000 SCROLLS.

TELEPORT NEWS brings you a rare COLOR PAPERBACK, right into current events that will literally advance the human race. An amazing ancient treasure map description has been found from around 12,500 years ago. The satillite Google Earth images, show that it appears to still be there.

Then we go right into ancient immortality. Can it really work? The ancient writings say so. Even if the ancient machine isn't found with the scrolls, it doesn't matter. The machine can be built right now with engineering testing. Are you a billionaire or millionaire and want to try to live a thousand years? Then build the machine. All the science, physicis, along with the ancient and current teleportation findings, say that a portal can be built. One opening to Heaven itself.

That higher frequency LIGHT is mentioned as the source for immortalily here on Earth.

Johnny Vincento is the world leading Applied physicist in THE TELEPORTATION PROJECT. Testing the Spacetime travel found in Thoth's ancient stone tablets for 20 years. The project had great success. He commented "Since that claim of Thoth's was correct, than everything else he says has to be true as well."

ISBN 9798370840517

90000

9 798370 840517

EINSTEIN - ROSEN BRIDGE NOW PROVEN: WORMHOLE ALL...

JOHNNY VINCENTO

Made in United States
Cleveland, OH
25 June 2025